This report was made possible by grants from the John D. and Catherine T. MacArthur Foundation in connection with its grant making initiative on Digital Media and Learning. For more information on the initiative visit www.macfound.org.

The John D. and Catherine T. MacArthur Foundation Reports on Digital Media and Learning

The Future of Learning Institutions in a Digital Age by Cathy N. Davidson and David Theo Goldberg with the assistance of Zoë Marie Jones

Living and Learning with New Media: Summary of Findings from the Digital Youth Project by Mizuko Ito, Heather Horst, Matteo Bittanti, danah boyd, Becky Herr-Stephenson, Patricia G. Lange, C. J. Pascoe, and Laura Robinson with Sonja Baumer, Rachel Cody, Dilan Mahendran, Katynka Z. Martínez, Dan Perkel, Christo Sims, and Lisa Tripp

Young People, Ethics, and the New Digital Media: A Synthesis from the GoodPlay Project by Carrie James with Katie Davis, Andrea Flores, John M. Francis, Lindsay Pettingill, Margaret Rundle, and Howard Gardner

Confronting the Challenges of Participatory Culture: Media Education for the 21st Century by Henry Jenkins (P.I.) with Ravi Purushotma, Margaret Weigel, Katie Clinton, and Alice J. Robison

The Civic Potential of Video Games by Joseph Kahne, Ellen Middaugh, and Chris Evans

The Future of Learning Institutions in a Digital Age

Cathy N. Davidson and David Theo Goldberg

With the assistance of Zoë Marie Jones

The MIT Press
Cambridge, Massachusetts
London, England

For information about special quantity discounts, please email special_sales@mitpress.mit.edu.

This book was set in Stone Serif and Stone Sans by the MIT Press.
Printed and bound in the United States of America.

Library of Congress Cataloging-in-Publication Data

Davidson, Cathy N., 1949–
The future of learning institutions in a digital age / Cathy N. Davidson and David Theo Goldberg ; with the assistance of Zoë Marie Jones.
p. cm.—(The John D. and Catherine T. MacArthur Foundation reports on digital media and learning)
Includes bibliographical references.
ISBN 978-0-262-51359-3 (pbk. : alk. paper)
1. Educational technology. 2. Internet in education. 3. Education—Effect of technological innovations on. 4. Educational change. 5. Organizational change. I. Goldberg, David Theo. II. Title.
LB1028.3.D27 2009 371.33′44678—dc22 2009007500

10 9 8 7 6 5 4 3 2 1

Contents

Series Foreword

The John D. and Catherine T. MacArthur Foundation Reports on Digital Media and Learning, published by the MIT Press, present findings from current research on how young people learn, play, socialize, and participate in civic life. The Reports result from research projects funded by the MacArthur Foundation as part of its $50 million initiative in digital media and learning. They are published openly online (as well as in print) in order to support broad dissemination and to stimulate further research in the field.

Acknowledgments

This has been a collective project from the beginning, and so our first acknowledgment goes to all those who supported and contributed. Funded by a grant from the John D. and Catherine T. MacArthur Foundation, as part of its initiative on Digital Media and Learning, we thank Constance M. Yowell, Director of Education in the MacArthur Foundation's Program on Human and Community Development, and the visionary behind the Digital Media and Learning Initiative. This project would not have been conceived, let alone possible, without her sustained and sustaining support. We also thank the President of the MacArthur Foundation, Jonathan F. Fanton, Vice President Julia Stasch, and our Program Officers for MacArthur's Digital Media and Learning Initiative, Craig Wacker and Ben Stokes. Michael Carter has been key also in seeing us through the final stages to publication, and we are enormously grateful for this.

Zoë Marie Jones, a doctoral candidate in the Department of Art, Art History, and Visual Studies at Duke University, joined this project in the Fall of 2007. Collaboration (especially with so many participants) is an enormous amount of work, and Ms. Jones has taken charge of the complicated process of integrating the responses and feedback from the virtual contributors into our final draft in a way that allows us to acknowledge those contributions, individually (where appropriate) and collectively. She has helped us to organize all of this input in a coherent fashion. She has also found models, resources, and additional bibliography sources that add practical wisdom and examples to our theoretical discussion of virtual institutions. We cannot thank her enough for her profound contributions, sustained insight, and dedication through the final months of this project.

Likewise, we are grateful to all those at the Institute for the Future of the Book for hosting our collaborative project, notably Bob Stein and Ben Vershbow.

Without the energetic participation of all of those who contributed to the Future of the Book collaborative site, to the HASTAC network and conference, and to our various forums on digital institutions, this book would look very different and, certainly, less visionary. The names of all who participated in this project are listed below.

We thank Duke University and the University of California and their administrators for having faith in HASTAC back in 2002, when many were skeptical that a virtual organization could have any impact or staying power.

At Duke University, our infrastructure comes from the John Hope Franklin Center for Interdisciplinary and International Studies and the John Hope Franklin Humanities Institute. For their early and continuing support, we thank: Provost Peter

Lange, Vice Provosts Gilbert Merkx and Susan Roth, Deans George McLendon, Gregson Davis, and Sarah Deutsch, and Franklin Humanities Institute Director Srinivas Aravamudan. We also thank, for their tireless contributions to HASTAC and this project, Mandy Dailey, Jason Doty, Erin Ennis, Sheryl Grant, Erin Gentry Lamb, Mark Olson, Jonathan E. Tarr, and Brett Walters.

At the University of California, our infrastructure comes from the University of California's Humanities Research Institute (UCHRI), the humanities institute for all of the universities in the California system. We thank key staff at UCHRI whose tireless contributions have made this project possible, most notably Suzy Beemer, Irena Richter, Jennifer Wilkens, Khai Tang, and Shane Depner.

The Future of Learning Institutions in a Digital Age

Overview of a Collaborative Project

This John D. and Catherine T. MacArthur Foundation Report is a redaction of the argument in our book-in-progress, currently titled *The Future of Thinking: Learning Institutions in a Digital Age*. That book, to be published in 2010, is merely the concrete (paper and online) manifestation and culmination of a long, complex process that brought together dozens of collaborators, face to face and virtually. The focus of all of this intense interchange was the shape and future of learning institutions. Our charge was to accept the challenge of an Information Age and acknowledge, at the conceptual as well as at the methodological level, the responsibilities of learning at an epistemic moment when *learning itself* is the most dramatic medium of that change. Technology, we insist, is not what constitutes the revolutionary nature of this exciting moment. It is, rather, the potential for shared and interactive learning that Tim Berners-Lee and other pioneers of the Internet built into its structure, its organization, its model of governance and sustainability.

This is an idealistic claim about the primacy of learning. We argue that the single most important characteristic of the

Internet is its capacity to allow for a worldwide community and its endlessly myriad subsets to exchange ideas, to learn from one another in a way not previously available. We contend that the future of learning institutions *demands* a deep, epistemological appreciation of the profundity of what the Internet offers humanity as a model of a learning institution.

To initiate and exemplify this rethinking of virtually enabled and enhanced learning institutions, we used this project to examine potential new models of digital learning. This project, in short, is an experiment. We engaged multiple forms of participatory learning to test the power of "virtual institutions" and to model other ways that virtual, digital institutions can be used for learning. The process itself informed every step of our thinking about new forms of alliances, intellectual networks, and comparative modes of interaction (digital or face-to-face) in a range of learning environments.

We are at an early and fast-changing moment in the development of online collaborative forms. We consider this report to be both a guide to others who may wish to pursue such a course and a historical record of a form that, we suspect, will continue to evolve as dramatically in the next five years as it has in the previous. Wikipedia, the world's most ambitious collaborative learning site to date, was after all launched only in 2001. Ours is by no means the first project to be written using collaborative tools, but it is among the first to consider a participatory, digital site from an institutional perspective and to incorporate an analysis of the process as part of its own research agenda.

As a writing exercise, our project is analogous to experiments in such recent books as Chris Anderson's *The Long Tail* and

MacKenzie Wark's *Gamer Theory*. Where our project differs in some respects from these and others is that it uses this experiment in participatory writing as a test case for virtual institutions, learning institutions, and a new form of virtual collaborative authorship. The names of all participants in this project are included in the appendix, and we consider their participation in this endeavor to be part of the content and method of the research itself.

The Future of Learning Institutions in a Digital Age began as a draft that we wrote together and then posted on a collaborative Web site developed by the Institute for the Future of the Book (http://www.futureofthebook.org) in January of 2007. The draft remained on the Institute's site for over a year (and still remains there) inviting comments by anyone registered to the site. An innovative digital tool, called Commentpress, allowed any reader to open a comment box for any paragraph of the text and to type in a response, and then allowed subsequent readers to add additional comments. Literally hundreds of viewers read the draft and dozens offered insights and also engaged in discussions with us or with other commentators.

We also held three public forums on the draft, including one at the first international conference convened by HASTAC ("haystack"), an acronym for Humanities, Arts, Science, and Technology Advanced Collaboratory.[1] HASTAC is a virtual network of academics and other interested educators in all fields who are committed to three principles fundamental to the future of learning institutions: first, the creative use and development of new technologies for learning and research; second, critical understanding of the role of new media in life, learning, and society; and third, pedagogical advancement of the goals of participatory learning. HASTAC is both the organizing collective body around which this monograph has developed and the centerpiece of our own commitment to virtual institutions.

This report points to only some of our conclusions about and principles for the future of learning institutions in a digital age. The full-length book goes much further. It offers pedagogical comparisons for teaching in new environments, detailing both the supports and inherent obstacles in collaborative teaching in virtual environments at this transitional moment. It theorizes what institutions are and how virtuality changes some institutional arrangements while requiring even stronger foundational support from traditional institutions in other ways. It re-theorizes the nature of learning and reconceives the concept of the institution as a mobilizing network resulting from the age of virtuality. It describes one such virtual institution—our own, HASTAC—in historical and institutional terms as a model of other such learning institutions. And it points to promises, problems, and even perils in the future of virtual learning institutions.

In addition, the longer book version includes extensive bibliographies to aid readers in their own endeavors to create learning institutions on new participatory models and offers a bibliography of models and examples of pioneering institutions that already are making the first steps at creating new learning networks.[2] In the online version of our book, URLs will point one directly to sites where one can find out more about a number of innovative participatory learning experiments and institutions. Although the scope of our main discussion is on university education and digital communities among adults, we have also included in our bibliography an annotated listing of K–12 and youth-oriented institutions that are taking the lead in exploring what virtual learning institutions might accomplish and how.

As authors, scholars, teachers, and administrators, we are part of many institutions. One conclusion we offer is that most virtual institutions are, in fact, supported by a host of real institutions and real individuals. We underscore this because it is part of the mythology of technology that its virtues, vitality, and value are "free." We seek to deflate that myth by underscoring how much the most inventive virtual and collaborative networks are supported by endless amounts of organization, leadership, and funding. Like a proverbial iceberg, sometimes the "free" and "open" tip of virtual institutions is what we see, but it is the unseen portion below the virtual waterline that provides the support. HASTAC, for example, could not exist without the tireless work of many individuals who contributed their time and energy. Those individuals are largely located at the two institutions that have provided the infrastructural support for HASTAC from its inception: Duke University and the University of California. It would not—could not—have gotten off the ground, survived, or thrived without that institutional support, as is the case, we insist, with any comparable virtual institution, at least at this moment in time.

There is a politics implicit in our emphasis on the infrastructural, leadership, organizational, and monetary costs beneath the tip of the information iceberg. The rhetoric of the "free" and "open" Internet is inspiring, and we heartily endorse open code and share-for-share not-for-profit licensing of the kind exemplified by Creative Commons. However, the digital divide still very much exists, across affluent countries such as the United States and throughout the wealthiest nations in Europe, and, with even greater disparity, across Third World countries. Bharat Mehra succinctly defines "digital divide" as "the troubling gap between those who use computers and the Internet and those who do not."[3] It is troubling—and more so. It is tragic, given how much of our global knowledge and commercial economy depend on mobile access of one form or another. To fail to acknowledge the cost of human labor and the amount of support necessary to sustain virtual institutions (as with traditional ones) obscures the importance of the extreme and ever-

access age associate book california center
change collaborative comments community
culture department digital director duke
education forms future hastac http
humanities information institutions
interactive internet knowledge learning
media models networked online open
participatory possibilities practices professor
project research school site social
students studies technologies think traditional
university virtual wikipedia work world

increasing distribution of wealth worldwide. There is also an extreme and, many argue, increasing (not decreasing) distribution of participation in the digital age.

We thank the institutions that support our research and our virtual endeavors. They are listed in the acknowledgments section of this report. Without their vision and commitment to this larger project of envisioning the best modes of learning for a digital age, this research project would not exist. It is not our purpose to condemn traditional institutions but, we fervently hope, to be among those inspiring the kinds of change that will make our learning institutions better suited to the experiences, skills, goals, and ambitions of the young people they serve and who will be responsible for shaping the future.

The Classroom or the World Wide Web? Imagining the Future of Learning Institutions in a Digital Age

The Classroom and the World Wide Web

Modes of learning have changed dramatically over the past two decades—our sources of information, the ways we exchange and interact with information, how information informs and shapes us. But our schools—how we teach, where we teach, who we teach, who teaches, who administers, and who services—have changed mostly around the edges. The fundamental aspects of learning institutions remain remarkably familiar and have done so for something like two hundred years or more. Ichabod Crane, that parody of bad teaching in Washington Irving's classic short story "The Legend of Sleepy Hollow" (1820), could walk into most college classrooms today and know exactly where to stand and how to address his class.

If we are going to imagine new learning institutions that are not based on the contiguity of time and place—*virtual* institutions—we have to ask, what are those institutions and what work do they perform? What does a virtual learning institution look like, who supports it, what does it do? We know that informal learning happens, constantly and in many new ways, because of the collaborative opportunities offered by social networking sites, wikis, blogs, and many other interactive digital sources. But beneath these sites are networks and, sometimes, organizations dedicated to their efficiency and sustainability. What is the institutional basis for their persistence? If a virtual site spans many individuals and institutions, who or what supports (in practical terms) the virtual site and by what mechanisms?

Our argument here is that our institutions of learning have changed far more slowly than the modes of inventive, collaborative, participatory learning offered by the Internet and an array of contemporary mobile technologies. Part of the reason for the relatively slow change is that many of our traditional institutions have been tremendously successful, if measured in terms of endurance and stability. It is often noted that, of all existing institutions in the West, higher education is one of most enduring. Oxford University, the longest continuously running university in the English-speaking world, was founded in the twelfth century.[4] Only the Catholic Church has been around longer and, like the Catholic Church, universities today bear a striking structural resemblance to what they were in medieval times. As is typically the case in the present, the medieval university was a separate, designated, physical location where young adults (students) came to be taught by those, usually older and more experienced, who were authorized (scholars, professors, dons) to impart their special knowledge, chiefly by lecturing. Over the years, such features as dormitories, colleges, and, later, departments were added to this *universitas* (corporation). The tendency toward increasing specialization, isolation, departmentalization, and advanced (graduate and professional school) training developed in the wake of the Enlightenment, gathering steam through the nineteenth and into the twentieth century.

Given this history, it is certainly hard to fathom something as dispersed, decentralized, and virtual as the Internet being a learning institution in any way comparable to, say, Oxford. We know, given these long histories, what a learning institution is—or we think we do. But what happens when, rivaling formal educational systems, there are also many virtual sites where learning is happening? From young kids customizing Pokémon (and learning to read, code, and use digital editing tools), to college students contributing to Wikipedia, to adults exchanging information about travel, restaurants, or housing via collaborative sites, learning is happening online, all the time, and in numbers far outstripping actual registrants in actual schools. What's more, they challenge our traditional institutions on almost every level: hierarchy of teacher and student, credentialing, ranking, disciplinary divides, segregation of "high" versus "low" culture, restriction of admission to those considered worthy of admission, and so forth. We would by no means argue that access to these Internet sites is equal and open worldwide (given the necessity of bandwidth and other infrastructure far from universally available as well as issues of censorship in specific countries). But there is certainly a greater degree of fluidity and access to participation than at traditional educational institutions.[5] So we re-ask our question: Are these Internet sites "learning institutions"? And, if so, what do these institutions tell us about the more traditional learning institutions such as schools, universities, graduate schools?

One of the best examples of a virtual learning institution in our era is Wikipedia, the largest encyclopedia compiled in human history and one "written collaboratively by volunteers from all around the world."[6] Sustaining Wikipedia is the Wikimedia Foundation, Inc., with its staid organizational charts and well-defined legal structures. What is the relationship between the quite traditional nonprofit corporation headquartered in San Francisco and the free, open, multilingual, online, global community of volunteers? Is the "institution" the sustaining organization, the astonishing virtual community, or the online encyclopedia itself?

When considering the future of learning institutions in a digital age, it is also important to look at the ways that digitality works to cross the boundaries within and across traditional learning institutions. How do collaborative, interdisciplinary, multi-institutional learning spaces help to transform traditional learning institutions and, specifically, universities? For example, how are the hierarchies of expertise—the ranks of the professoriate and also the divide of undergraduates, graduate students, and faculty (including adjunct faculty, tenure-track junior faculty, tenured, distinguished, and emeriti faculty)—supported and also undermined by new digital possibilities? Are there collaborative modes of participatory learning that help to rethink traditional pedagogical methods? And what might learning institutions look like—what *should* they look like—given the digital potentialities and pitfalls at hand today?

We are concerned to conjecture about the character of learning institutions and how they change, how they change those who belong to them, and how people can work together to change them. Our primary focus is higher education. It is daunting to think that universities have existed in the West since medieval times and in forms remarkably similar to the universities that exist today. Will they endure for hundreds of years more even as learning increasingly happens virtually, globally, and collaboratively? It is our hope that thinking about the potential of new ways of knowing might inspire the revitalization of those institutions of advanced formal learning.

Participatory Learning

A key term in thinking about these emergent shifts is *participatory learning*. Participatory learning includes the many ways that learners (of any age) use new technologies to participate in virtual communities where they share ideas, comment on one another's projects, and plan, design, implement, advance, or simply discuss their practices, goals, and ideas together.

This method of learning has been promoted both by HASTAC and by the John D. and Catherine T. MacArthur Foundation's Digital Media and Learning Initiative. Participatory learning begins from the premise that new technologies are changing how people of all ages learn, play, socialize, exercise judgment, and engage in civic life. Learning environments—peers, family, and social institutions (such as schools, community centers, libraries, museums, even the playground, and so on)—are changing as well. The concept of participatory learning is very different from "IT" (Instructional Technology). IT is usually a toolkit application that is predetermined and even institutionalized with little, if any, user discretion, choice, or leverage. IT tends to be top-down, designer determined, administratively driven, commercially fashioned. In participatory learning, outcomes are typically customizable by the participants.

Since the current generation of college student has no memory of the historical moment before the advent of the Internet, we are suggesting that participatory learning as a practice is no longer exotic or new but a commonplace way of socializing and learning. For many, it seems entirely unremarkable.[7] Global business more and more relies on collaborative practices where content is accretive, distributed, and participatory. In other areas too—from the arts to the natural and computational sciences and engineering—more and more research is being enacted collaboratively. A *New York Times* article from 2008 even suggested that a future Nobel Prize winner might not be an oncology researcher at a distinguished university but a blogging community where multiple authors, some with no official form of expertise, actually discover a cure for a form of cancer through their collaborative process of combining, probing, and developing insights online together.[8]

Participatory learning is happening now—not in the future, but now. Those coming into our educational system rely on participatory learning for information about virtually everything in their lives. Adults, too, turn first to the Internet and the "wisdom of crowds" and "smart mobs" to help them make decisions about which car to buy, which cell phone service to use, which restaurants to frequent, and even which form of heart surgery promises the best results with the least risk. Business and other professions turn more and more to collaborative learning forms. Again, this is not the future. This is the condition of life now, in 2009, for a majority certainly in the global north but increasingly through the use of mobile technologies in the global south, too.

This puts education and educators in the position of bringing up the rearguard, of holding desperately to the fragments of an educational system which, in its form, content, and assessments, is deeply rooted in an antiquated mode of learning. Every university in the global north, of course, is spending large sums of money revamping its technology offerings, creating great wired spaces where all forms of media can be accessed from the classroom. But how many have actually rethought the modes of organization, the structures of knowledge, and the relationships between and among groups of students, faculty, and others across campus or around the world? That larger challenge—to harness and focus the participatory learning methods in which our students are so accomplished—is only now beginning to be introduced and typically in relatively rare and isolated formats.

Most university education, certainly, is founded on ideas of individual training, discrete disciplines, and isolated achievement and accomplishment. What we want to ask is how much this very paradigm of individual achievement supports the effective learning styles of today's youth and prepares them for increasingly connected forms of civic participation and global commerce—or how much it is at odds with contemporary culture. That needs to be stated more forcefully: The future of conventional learning institutions is past—*it's over*—unless those directing the course of our learning institutions realize, now and urgently, the necessity of fundamental and foundational change.

Most fundamental to such a change is the understanding that participatory learning is about a process and not always a final product. We are concerned here not just with a prognostication about future institutions for learning, but with considering, even with projecting, how learning happens *today*—not in some distant utopian or dystopian future.

As noted above, we posted an early draft of this essay on Commentpress, the Web-based tool developed by the Institute for the Future of the Book as a variation of the blogging software, Wordpress. Released in 2007, Commentpress allows an online text to be "marked up" in a digital version of margin notes. In doing so, we made authorship a shared and interactive experience, in which we were able to engage in online conversation with those reading and responding to our work.[9] That is a version of participatory learning.

Box 1

The Institute for the Future of the Book

Our tools of learning are shifting increasingly from the printed page to digital media. The Institute for the Future of the Book (http://www.futureofthebook.org/) takes as its mission the chronicling of this shift and the development of digital resources to promote innovative reimaginings of the book.

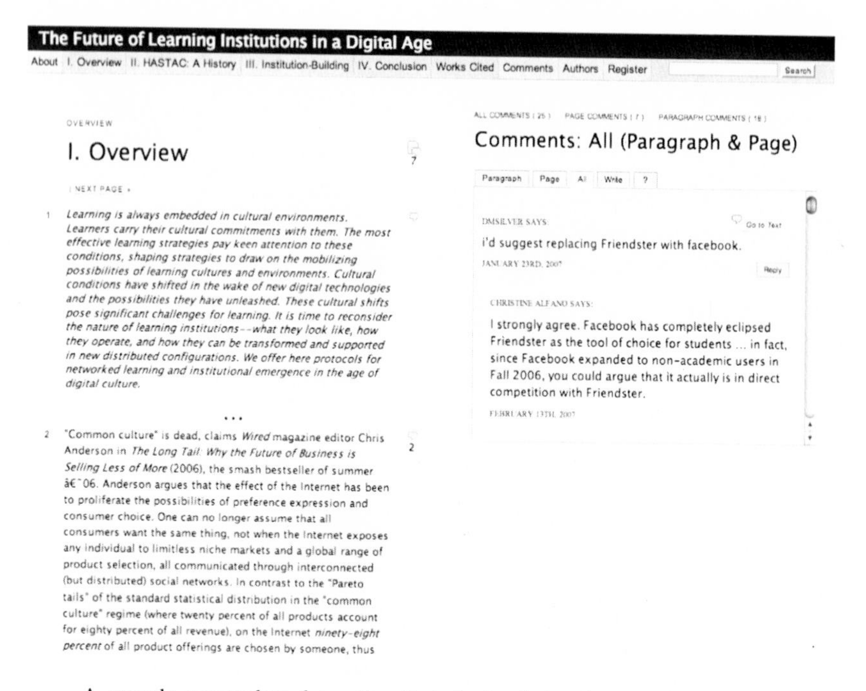

A sample screenshot from the first draft of the Future of Learning Project

Participatory learning is not simply about interaction (we all have plenty of that in our lives) but of interaction that, because of issues of access, means that one is co-creating with myriad people who are strangers and who can remain anonymous. People can respond candidly. From such a process, one learns and continues to learn from others met (if at all) only virtually, whose institutional status and credentials may be unknown.

With participatory learning, the play between technology, composer, and audience is no longer passive. Indeed, participatory learning blurs these traditional lines.[10] That blurring raises important questions for those trained as academics under a different set of premises. In the current academy, virtually everything in a scholar's life is based on peer review and institutionally ordained authority. Who counts as a "peer" is carefully defined and which institution counts as a peer institution is also certified through accreditation and other ranking practices. Within those sets of rules, the lines of authorship and authority are clearly delineated. There is a hierarchy within the professoriate that models a similar hierarchy of teacher and student. With participatory learning these conventional modes of authority break down. In our own case of composing the first outline drafts of this document, we found the process of making our ideas accessible to anyone who wished to comment to be sometimes frustrating, occasionally embarrassing (no one likes to be called out in public), but, in the end, intellectually exhilarating in both content and in form.

Remix Authorship

As often happens in the history of technology, a significant new device—such as a breakthrough in new hardware or software—has an impact on a variety of the social and political conditions around it. These impacts may be large or small, general or local. On the more local end of the spectrum, in the case of the Commentpress tool, our concept of publishing changed, as did our concept of authorship over the course of our experiment with collective feedback and revision. Is the first or the final version of our text the "published" version of the essay?

Obviously the answer has to be "both." The concept of authorship (a subject to which we will return) also needs to be reassessed because of the interactive publishing process.[11] As any historian of the book knows, you cannot change one part of the publishing circuit without shifting the dynamics among all the other parts of the process. All aspects of publishing are interconnected, from the materiality of production (books or Web sites) to distribution networks (bookstores or downloads), to readership and even such foundational concepts as "literacy" (a term that, with participatory learning, must include digital literacy).[12] These affect content, too.

The implications of the Commentpress interaction are fascinating for thinking about the future of learning institutions. Anyone could join the Commentpress Web site and make notes on our report, without the benefit of any specific institutional membership. As long as they heard about the project from some source—*networking* is another crucial component of participatory learning that we will return to later—they could register and comment. Given that one could log on from an Internet cafe in Thailand or from a graduate research program in Boston, this process offered important issues of access, authority, and anonymity. It also offered the retreat, if not the vanishing altogether, of traditional institutional structures and implicit notions of institutional membership and hierarchy marking most forms of feedback to scholarly work—such as shared membership in a classroom, an academic department, or a professional association. Participation required only access to a computer and enough literacy to be able to read, comprehend, respond to, and influence what we wrote. In many ways, the

Institute for the Future of the Book is an extension of the first subscription libraries. Ben Franklin established the Library Company of Philadelphia in 1731 to give more readers access to more kinds of knowledge. The Internet, surely, has redefined access (and its limits) for the twenty-first century. It has also dramatically reordered, if not undermined, traditional hierarchical orders of knowledge authority based on domain expertise sanctioned by institutional license.

Indeed, in a magisterial article about new technologies, the distinguished historian of the book (and now the director of Harvard University's Libraries), Robert Darnton, has suggested that we live in not the first but actually the fourth great Information Age. The previous ones he defines as the invention of writing in approximately 4000 BC, the turn from the scroll to the codex in the third century AD, the invention of the printing press (by the Chinese in 1045 and in the West, by Gutenberg, in 1450), and now, the invention of the Internet. Of all of these, Darnton argues, the Internet has had the fastest and the most geographically extensive effect on every aspect of knowledge making and all of the arrangements of life around how we make, exchange, share, correct, and publish our ideas.[13] It has also shifted both the perception and the reality of who makes knowledge, how it is authorized and legitimated.

And yet, to return to our central point, our learning *institutions*, for the most part, are acting as if the world has not suddenly, irrevocably, cataclysmically, epistemically changed—and changed precisely in the area of learning. We are not clear if this is so much an ostrich time for learning institutions or (to use a different animal metaphor) a deer-in-the-headlights time.

In any case, most institutions are stuck in an epistemological model of the past, even as they pour tens or even hundreds of thousands of dollars into IT that promises a technological future. Yet, we are no longer talking about the future. Institutional change is happening as we write.

The Challenge

No one (except perhaps politicians) promotes change for the sake of change. Implicit in a sincere plea for transformation is an awareness that a current situation needs improvement. When we advocate institutional change for learning institutions, we are making assumptions about the deep structure of learning, about cognition, about the way youth today learn (about) their world in informal settings, and about a mismatch between the excitement generated by informal learning and the routinization of learning so common to many of our institutions of formal education. We advocate institutional change because we believe our current formal educational institutions are not taking enough advantage of the modes of digital and participatory learning available to students today.

Youth who learn via peer-to-peer mediated forms may be less likely to be excited and motivated by the typical forms of learning than they were even a decade ago. Too many conventional modes of learning tend to be passive, lecture driven, hierarchical, and largely unidirectional from instructor to student. As Wheat (logon name) notes on the Institute of the Future of the Book, "open-ended assignments provide the opportunity for creative, research-based learning."[14] And yet in the vast majority of formal educational settings, partly as a concomitant of cutbacks to education resulting in increased class size but also partly a function of contemporary reified culture, the multiple-choice test has replaced the research paper or more robustly creative group-produced projects.

On the K–12 level (primary and secondary public schools), governmentally mandated programs, including those such as "No Child Left Behind," tend overwhelmingly to reinforce a form of one-size-fits-all education, based on standardized testing. Call this cloned learning, cloning knowledge, and clones as the desired product. Such learning models—or "cloning cultures"—are often stultifying and counter-productive, leaving many children bored, frustrated, and unmotivated to learn.[15]

The deplorable U.S. high school dropout rates now amount to close to 35 percent of those who begin public schools in the United States.[16] Of special urgency is the surging gap between the wealthy and the poor, a gap that correlates in both directions with educational levels.[17] Youth from impoverished backgrounds are statistically most likely to drop out of school; high school dropouts earn less than those with a diploma, and significantly less again than those with a university degree. Incarceration rates, which have soared more than tenfold since 1970, also correlate closely with educational failure and impoverishment. Seventy-five percent of those imprisoned tend to be illiterate, earning under $10,000 per year at the time of arrest.[18] Currently, according to Human Rights Watch, the United States has the highest incarceration rate of any nation on earth, higher even than China, with 762 of every 100,000 U.S. residents currently in jail (as compared to incarceration rates in the United

Kingdom of 152 per 100,000 residents, and, in Canada and France, 108 and 91, respectively).[19]

In the United States, incarceration correlates with poverty and digital access correlates with educational opportunity and wealth. Despite government pronouncements to the contrary, "digital divide" is not just an old concept but a current reality.[20] Access to computers remains unevenly distributed. In our comments about formal education, implicit is an awareness that even the most basic resources (including computers) are lacking in the nation's most impoverished public schools as well as in the nation's poorest homes.

Wealth, formal education, race, and gender are important interacting factors in the certification of what constitutes "merit" and "quality." Nevertheless, and even granting the digital divide, there is a generational shift in the kinds of learning happening by those both living above the poverty line and those more impoverished youth accessing such media in perhaps more limited form (often through community centers and libraries). An increasing number of those born after 1983 (the desktop) and 1991 (the Internet) learn through peer-to-peer knowledge networks, collaborative networks, and aggregated private and open source social spaces (from MySpace and Facebook to del.icio.us).

Given that the entering college class was born in 1989 or 1990, we are talking about a cultural change that touches every aspect of the educational system as well as nonformal learning environments for all ages. The so-called "Millennials" are, in fact, not the only age group being transformed by digital technologies. We note in passing that the average age of a *World of Warcraft* game player is 28.[21]

Because of the Internet, more and more choices are available to the public, in everything from consumer products to software, social networks, modes of play, knowledge/data repositories, and cultural archives. While contributors to Wikipedia are just 5 percent of users, that is a far larger contributing base than to traditional encyclopedias, and many more consult Wikipedia on a daily basis than they ever consulted print-based encyclopedias. Learning, too, has a "long tail," where more and more is available virtually, to potentially much wider, more distributed, and diverse ranges of people.

We do not claim to have solutions for these massively complex social issues, nor do we claim to understand fully the relationships between and among the various developments we have listed. However, we do believe the opportunity now exists to mobilize educators and learners to more energetic and productive learning ends. Interactive technologies and collaborative learning have inspired enormous excitement, and contemporary youth exhibit great facility in negotiating the use of new media. We believe, accordingly, that learning institutions can be developed to do a better job of enlisting the imagination of youth and to use the specialized interests of young people for the purposes of placing in practice wise and rigorous forms of knowledge sharing.

To accomplish this end will require that educators rethink their most cherished methodologies and assumptions. It is not easy to rethink knowledge in the Net Age.[22] As open source legal theorist and activist James Boyle notes in his witty and terse article "A Closed Mind about an Open World," we have been conditioned by a confluence of factors, economic and social, political and cultural, to acquire an "openness aversion."[23] The familiar is safe, easy, reliable. Boyle suggests that aversion to openness—to be disposed against the challenge of the unforeseen—is an actual cognitive bias that leads us to "undervalue the importance, viability and productive power of open systems, open networks and non-proprietary production." To overcome this bias requires that knowledge producers (all of us involved in the practices of teaching, in whatever current institutional configuration) rethink every aspect (from economic theory to citation form) of what we think of as "knowledge production."

Digital Presence and Digital Futures

Digital technologies increasingly enable and encourage social networking and interactive, collaborative engagements, including those implicating and impacting learning. And yet traditional learning institutions, whether K–12 or institutions of higher learning, continue to privilege individualized performance in assessments and reward structures. Born and matured out of a century and a half of institutional shaping, maturing, and hardening, these assessment and reward structures have become fixed in place. But they now serve also to weigh down and impede new learning possibilities.

Digital technologies have dramatically encouraged self-learning. Web interfaces have made for less hierarchical and more horizontal modes of access. The Web has also facilitated the proliferation of information, from the inane and banal to the esoteric and profound, from the patently false, misleading, even (potentially) dangerous and destructive to the compelling, important, and (potentially) life-enhancing and life-saving. But the relative horizontality of access to the Web has had another surprising effect: it has flattened out contributions to knowledge making, too, making them much less the function of a credentialed elite and increasingly collaboratively created.

What are the implications of this dual horizontality—of access and contribution—for learning, then? It is to that question we turn next.

Pillars of Institutional Pedagogy: Ten Principles for the Future of Learning

We suggest that the following ten principles are foundational to rethinking the future of learning institutions.[24] We see these principles as riders, both as challenges and as the general grounds on which to develop creative learning practices, both transformative and transforming as new challenges emerge and new technological possibilities are fashioned.

1. Self-Learning

Self-learning has bloomed; discovering online possibilities is a skill now developed from early childhood through advanced adult life. Even online reading, as Alan Liu reminds us, has become collaborative, interactive, nonlinear and relational, engaging multiple voices.[25] We browse, scan, connect in mid-paragraph if not mid-sentence to related material, look up information relevant or related to what we are reading. Sometimes this mode of relational reading might draw us completely away from the original text, hypertextually streaming us into completely new threads and pathways across the information highways and byways. It is not for nothing that the Internet is called the "Web," sometimes resembling a maze but more often than not serving as a productive if complex and challenging switchboard.

2. Horizontal Structures

Relatedly, an increasingly horizontal structure of learning puts pressure on how learning institutions—schools, colleges, universities, and their surrounding support apparatuses—enable learning. Institutional education has tended to be authoritative, top-down, standardized, and predicated on individuated assessment measured on standard tests. Increasingly today, work regimes involve collaboration with colleagues in teams. Multitasking and overlapping but not discrete strengths and skills reinforce capacities to work around problems, work out solutions, and work together to complete projects. Given the range and volume of information available and the ubiquity of access to information sources and resources, learning strategy shifts from a focus on information as such to judgment concerning reliable information, from memorizing information to how to find reliable sources. In short, from learning *that* to learning *how*, from content to process.

3. From Presumed Authority to Collective Credibility

Learning is shifting from issues of authoritativeness to issues of credibility. A major part of the future of learning is in developing methods, often communal, for distinguishing good knowledge sources from those that are questionable. Increasingly, learning is about how to make wise choices—epistemologically, methodologically, concerning productive collaborative partnerships to broach complex challenges and problems. Learning increasingly concerns not only how to resolve issues regarding information architecture, interoperability and compatibility, scalability and sustainability, but also how to address ethical dilemmas. It concerns, in addition, issues of judgment in resolving tensions between different points of view in increasingly interdisciplinary environments. We find ourselves increasingly being moved to interdisciplinary and collaborative knowledge-creating and learning environments in order to address objects of analysis and research problems that are multidimensional and complex, and the resolution of which cannot be fashioned by any single discipline. Knowledge formation and learning today thus pose more acute challenges of trust. If older, more traditional learning environments were about trusting knowledge authorities or certified experts, that model can no longer withstand the growing complexities—the relational constitution of knowledge domains and the problems they pose.

4. A De-Centered Pedagogy

In secondary schools and higher education, many administrators and individual teachers have been moved to limit use of collectively and collaboratively crafted knowledge sources, most notably Wikipedia, for course assignments or to issue quite stringent guidelines for their consultation and reference.[26] This is a catastrophically anti-intellectual reaction to a knowledge-making, global phenomenon of epic proportions.

To ban sources such as Wikipedia is to miss the importance of a collaborative, knowledge-making impulse in humans who are willing to contribute, correct, and collect information without remuneration: by definition, this *is* education. To miss how much such collaborative, participatory learning underscores the foundations of learning is defeatist, unimaginative, even self-destructive.[27]

Instead, leaders at learning institutions need to adopt a more inductive, collective pedagogy that takes advantage of our era. John Seely Brown has noted that it took professional astronomers many years to realize that the benefits to their field of having tens of thousands of amateur stargazers reporting on celestial activity far outweighed the disadvantages of unreliability. This was a colossal observation, given that among the cohort of amateur astronomers were some who believed it was their duty to save the earth from Martians. In other words, professional astronomers had large issues of credibility that had to be counterpoised to the compelling issue of wanting to expand the knowledge base of observed celestial activity. In the end, it was thought that "kooks" would be sorted out through Web 2.0 participatory and corrective learning. The result has been a far greater knowledge, amassed in this participatory method, than anyone had ever dreamed possible, balanced by collective and professional procedures for sorting through the data for obviously wrong or misguided reportings. If professional astronomers can adopt such a de-centered method for assembling information, certainly college and high school teachers can develop a pedagogical method also based on collective checking, inquisitive skepticism, and group assessment.[28]

5. Networked Learning

Socially networked collaborative learning extends some of the most established practices, virtues, and dispositional habits of individualized learning. These include taking turns in speaking, posing questions, listening to and hearing others out. Networked learning, however, goes beyond these conversational rules to include correcting others, being open to being corrected oneself, and working together to fashion workarounds when straightforward solutions to problems or learning challenges are not forthcoming. It is not that individualized learning cannot end up encouraging such habits and practices. But they are not natural to individual learning, which leans on a social framework that stresses competition and hierarchy rather than cooperation, partnering, and mediation. If individualized learning is chained to a social vision prompted by "prisoner dilemma" rationality in which one cooperates only if it maximizes narrow self-interest, networked learning is committed to a vision of the social stressing cooperation, interactivity, mutuality, and social engagement for their own sakes and for the powerful productivity to which it more often than not leads. The power of ten working interactively will almost invariably outstrip the power of one looking to beat out the other nine.

6. Open Source Education

Networked learning is predicated on and deeply interwoven into the fabric of open source culture.[29] Open source culture seeks to share openly and freely in the creation of culture, in its production processes, and in its product, its content. It looks to have its processes and products improved through the contributions of others by being made freely available to all. If individualized learning is largely tethered to a social regime of copyright-protected intellectual property and privatized ownership, networked learning is committed in the end to an open source and open content social regime. Individualized learning tends overwhelmingly to be hierarchical: one learns from the teacher or expert, on the basis overwhelmingly of copyright-protected publications bearing the current status of knowledge. Networked learning is at least peer-to-peer and more robustly many-to-many.

In some circumstances, where resources are unevenly distributed, the network operates according to what we call a *many-to-multitudes model*. That is, a group that has access to resources sustains and supports the infrastructure required to engage in what are equitable *intellectual* exchanges with those who do not have the financial resources to sustain digital connection. Many international social movements—such as those focused on Darfur or Tibet—operate from this many-to-multitudes interactivity where financial resources on one end are balanced by local expertise and human investment and labor on the other for interchanges that are rich and socially valuable for all participants. Many-to-multitudes does not erase the digital divide but, rather, acknowledges its material reality and provides a more collective model of capital (monetary capital and human capital) to promote interchange. The desire (on all sides) for interactivity fuels this digitally driven form of social networking, as much in learning as in economic practices. It provides the circuits and nodes, the combustion energy and driving force for engaged and sustained innovative activity, sparking creativity, extending the circulation of ideas and practices, making available the test sites for innovative developments, even the laboratory for the valuable if sometimes painful lessons to be learned from failure.

7. Learning as Connectivity and Interactivity

The connectivities and interactivities made possible by digitally enabled social networking in its best outcomes produce learning ensembles in which the members both support and sustain, elicit from and expand on each other's learning inputs, contributions, and products. Challenges are not simply individually faced frustrations, Promethean mountains to climb alone, but mutually shared, to be redefined, solved, resolved, or worked around—together.

An application such as Live Mesh allows one to unite and synchronize one's entire range of devices and applications into a seamless web of interactivity. It enables instantaneous file- and data-sharing with other users with whom the user is remotely connected, thus allowing at least potentially for seamless and more or less instant communication across work and recreational environments. Our technological architecture thus is fast making net-*working*—in contrast with isolated, individualized working—the default. Slower to adapt, the organizational architecture of our educational institutions and pedagogical delivery are just starting to catch on and catch up.

8. Lifelong Learning

It has become obvious that from the point of view of participatory learning there is no finality. *Learning is lifelong.* It is lifelong not simply in the Socratic sense of it taking that long to realize that the more one knows the more one realizes how little one knows. It is lifelong in the sense also, perhaps anti-Platonically, that the increasingly rapid changes in the world's makeup mean that we must necessarily learn anew, acquiring new knowledge to face up to the challenges of novel conditions as we bear with us the lessons of adaptability, of applying lessons to unprecedented situations and challenges. It is not just that economic prospects demand it; increasingly "our" sociality and culture now do, too.

It remains an open question still whether connected, open source, interactive, networked, horizontal, lifelong learning will have a transformative epistemological impact on what we learn at our educational institutions. But what is certain is that the pedagogical changes we have enumerated have radically changed *how* we know how we *know*.[30]

9. Learning Institutions as Mobilizing Networks

Collaborative, networked learning alters also how we think about learning *institutions*, and network culture about how to conceive of institutions more generally. Traditionally, institutions have been thought about in terms of rules, regulations, norms governing interactivity, production, and distribution within the institutional structure. Network culture and associated learning practices and arrangements suggest that we think of institutions, especially those promoting learning, as mobilizing networks. The networks enable a mobilizing that stresses flexibility, interactivity, and outcome. And the mobilizing in turn encourages and enables networking interactivity that lasts as long as it is productive, opening up or giving way to new interacting networks as older ones ossify or newly emergent ones signal new possibilities. Institutional culture thus shifts from the weighty to the light, from the assertive to the enabling. With this new formation of institutional understanding and practice, the challenges we face concern such considerations as reliability and predictability alongside flexibility and innovation.

10. Flexible Scalability and Simulation

Networked learning both facilitates and must remain open to various scales of learning possibility, from the small and local to the widest and most far-reaching constituencies capable of productively contributing to a domain, subject matter, knowledge formation and creation. New technologies allow for small groups whose members are at physical distance to each other to learn collaboratively together and from each other; but they also enable larger, more anonymous yet equally productive interactions. They make it possible, through virtual simulations, to learn about large-scale processes, life systems, and social structures without either having to observe or recreate them in real life. The scale will be driven by the nature of the project or knowledge base, ranging from a small group of students working on a specific topic together to open-ended and open-sourced contributions to the Encyclopedia of Life or to Wikipedia. Learning institutions must be open to flexibility of scale at both ends of the spectrum, devising ways of acknowledging and rewarding appropriate participation in and contributions to such collective and collaborative efforts rather than too quickly dismissing them as easy or secondary or insufficiently individualistic to warrant merit.

Challenges from Past Practice, Moving Fast Forward

We have been stressing the range of opportunities and the transformative possibilities for learning at all levels as a result of readily available and emergent digital technologies. We are not naive, and we realize also the challenges, limitations, and misdirections—in short, the opportunity costs—resulting from these developments. Some of this comes inevitably with the unsettling of long-established ways of doing things. When well-established modes of making knowledge sediment, they can become at least restrictive, if not unproductive. As new modes challenge, the old institutional structures can either dig in and refuse to respond other than to dismiss, or they can seek to work out renewed and renewing regimes disposed to take advantage of the possible productive elements.

The challenges by participatory learning to institutional order in higher education (though these challenges count too in thinking about other institutional levels and formations) range from the banal to the constitutive, from the disciplining of behavioral breaches of protocol and expectation to normative conceptions of what constitutes knowledge and how it is authorized.

Conclusion: Yesterday's Tomorrow

We have been re-examining some of the key premises and the roles they have played in shaping learning institutions in general and higher education more particularly, especially since the end of World War II. Access to education at all levels for larger and larger segments of the population was crucial to settling class conflict and the development of middle class aspiration in the wake of the Great Depression. Publicly funded schools, community colleges, and technical training institutions as well as universities drew rapidly expanding numbers, shaping what it meant to be an educated citizen, a productive employee, and a moral person. As a consequence, income and wealth expanded from the 1930s to the 1980s, though significantly more so for some groups than for others. Demand for labor for the most part outstripped its supply, creating an upward spiral for wages and subsequent wealth and quality of life, in particular from one generation to the next.

All this began to change at the onset of the 1980s. The neoliberal cuts in state services, including notably to educational resources at all levels, driven in the past three decades by the marriage of political economy and the culture wars, have meant a resurgence in inequality tied to educational access, the insistence on test-driven pedagogy, and the class bifurcation, racially molded, in access to creative learning practices. The earlier emphasis on public education has given way to its privatizing erosion at all levels, whether through experiments in corporately run schools and school districts, through charter schools and vouchers, through distance learning programs for the racial poor on reservations, the dramatic privatization of higher education, or through the introduction of user fees for the likes of libraries and especially museums and their transformation by the cultural industry model of urban branding into sites for tourist attraction.[31]

No institution of higher education in the country today has tested in a comprehensive way new methods of learning based on peer-to-peer distributed systems of collaborative work characteristic of the new Internet age. We have mentioned earlier a couple of experimental examples emerging at the school level. Social psychologists such as Joshua Aronson and Claude M. Steele have established quite conclusively that collaborative learning is beneficial across class and culture, race and religion. These new modes of distributed, collaborative engagement are likely both to attract a broad range of motivated learning across conventional social divisions (think of the anonymous interactions across classes and races in online gaming) and to inspire new forms of knowledge and product creation. But can we really say, in 2009, that the *institutions* of learning—from preschool to the PhD—are suited to the new forms of learning made available by digital technologies? Is there an educational enterprise anywhere in the world redesigned with the deep assumptions of networked thinking core and central to its lesson planning? Has anyone yet put into institutional practice at the level of higher education what John Seely Brown is calling a "social life of learning for the 'Net age'"?[32]

If we face a future where every person has (easy access to) a laptop or networked mobile device, what will it mean? What will it mean for institutionally advocated, mediated, and activated learning? How will educators use these tools and this moment? How will users—learners—adapt them to learning functionality, access, and productive learning possibilities? Will what is learned and the new methods of learning alter as a consequence, becoming quicker but shallower, more instrumental and less reflective? Or will the social networking possibilities prompt greater reflexivity, a more sustained sociality in which the positions and concerns of the otherwise remote are more readily taken into consideration in decision making? How *can* we use these tools to inspire our most traditional institutions of learning to change?

It is to the illustration of what learning institutions currently offer that we next turn. In the book-length version we discuss at length also the obstacles learning institutions now traditionally pose to innovative learning that takes advantage of the online learning practices and possibilities available. It is our hope that, by assessing some of the institutional barriers as well as some of the institutional promise, we can begin to mobilize our institutions to envision formal, higher education as part of a continuum with (rather than a resistance to) the collaborative, participatory, networked engagements that our students participate in online today.

It would be easy to fall into handwringing, to say our institutions of education are antiquated and therefore doomed. In fact, their persistence suggests that, outmoded as they may be, they are not only not doomed—but thriving—at present. The baby boom of the baby boom, in 2009, makes admission to a college or university more competitive than it has ever been. A college degree is still the key to success, as all comparative studies of income levels and educational attainment attest. So it is our objective in this report not to simply dismiss, excoriate, or condemn, but to look at places where institutions *are* and *could be* productively changing in order to provide examples for those innovative educators, administrators, students, and parents who seek change and are not sure where to look for models.

MIT professor and digital learning pioneer Henry Jenkins has usefully spoken of the "convergence" resulting from networking a culture of new models and forms and contributions with older models. The convergence is not just the new working on and not simply around older forms, but thoroughly remixing and modding them, transforming them piecemeal, expanding and enlarging access to them.[33] So, too, we believe, is the charge and challenge to the immediate future of learning institutions. Remixed learning institutions may well be the model of the future.

Rather than describe that model in words, we offer here a portfolio of models, with URLs and screenshots from Web pages, of educational enterprises that are seeking to change not just the tools of the trade of education—but the trade itself. How successful these experiments in new institutional formations will be remains in question. We offer these concluding examples simply to provoke thought, not to foreclose it, to prod imaginations. In the book version we engage a more sustained account of the possibilities, challenges, and indeed failures posed by such examples. It is our hope that, in thinking together with all of those who have contributed to our forums—face-to-face at HASTAC gatherings as well as online with the Institute for the Future of the Book collaborative tool—we have begun a process, together, of envisioning better ways to think of the future of learning institutions in our digital age.

Appendix: Portfolio of Virtual Learning Institutions: Models, Experiments, and Examples to Learn and Build On

Gaming and Virtual Environments in Education

Not only is educational gaming becoming seen as a viable alternative to a formal education, but other types of virtual environments and Massively Multiplayer Online Games are being recognized for their educational components. Below are just some of the most popular examples of these educational alternatives.

Virtual Worlds

An undergraduate course, "Field Research Methods in Second Life," conducted entirely in the virtual world of Second Life, was taught by Ed Lamoureux of Bradley College in January 2007. Due to the success of this class, Lamoureux, Professor Beliveau in Second Life, has created two new courses based on the same principles, "Introduction to Field Research in Virtual Worlds" and "Field Research in Virtual Worlds."

a

b

(a) Bradley College professor Ed Lamoureux with his avatar, Professor Beliveau; and (b) student Ryan Cult with his avatar, Judge Canned.

Single-Player Computer Games

Classic computer games such as *SimCity* and *Civilization* are being given new life through their use in the classroom because of their ability to simulate complete environments. These games are often used to teach students about building and maintaining social and physical institutions.

a

b

Screen shots of (a) *SimCity* and (b) *Civilization*

Massively Multiplayer Online Games

Massively Multiplayer Online Games are attracting scholarly attention as an important social phenomenon. Games such as *World of Warcraft* offer alternative worlds where social functions, learning, and the development of social, tactical, and work skills can be practiced in a virtual environment. Researchers are also beginning to look at these games as a way to study model societies and social interactions.

Screenshot from *World of Warcraft*

Serious (or Educational) Games

Gamelab Institute of Play The Gamelab Institute of Play (http://www.instituteofplay.com) promotes gaming literacy (which they define as "the play, analysis, and creation of games") as a foundation for learning, innovation, and change in a digital society. Although they have been involved in several initiatives that target teenagers, such as the Gaming School, they offer

institute of play
PLAYBOOK
ABOUT
TEAM
PRESS
VOLUNTEER
CONTACT

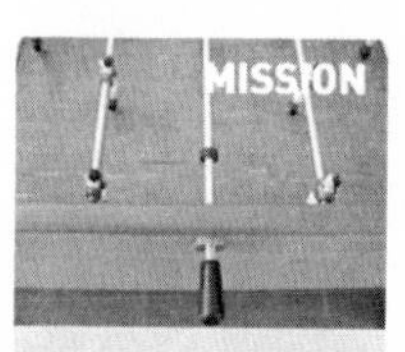
MISSION
We promote GAMING LITERACY: the play, analysis, and creation of games, as a foundation for learning, innovation, and change in the 21st century. Through a variety of programs centered on game design, the Institute engages audiences of all ages, exploring new ways to think, act, and speak through gaming in a social world.
Jump to...
Gamestar Mechanic Goes to Minneapolis
Posted December 10th, 2007 by Katie
in arena

PLAY
viddler
00:03:33
MENU
For a group of teens attending the Design Camp at the University of Minnesota in August, seeing the world as a potential game space took on new meaning. Eric Socolofsky and I--two members of the Gamestar Mechanic design team (a game developed by Gamelab--spent the week working with the teens to design and playtest games in the alpha build of the editor we currently have for the game.
ARENA
The Game School
MiLK
Pilots & Toolkits
Research
ON DECK
Summer Teacher Institute
Being, Space and Place
Texterritory
Being Me
FAQs
What are gaming literacies?
What kinds of games?
Do you work only with schools?
When will the school open?
Do you make games?
Can I come work with you?
Do you only work with youth?
Can I use your research?
Do you do consulting?
Who is on your Board?
Who is Gamelab?
LINKS

The Gamelab Institute of Play (http://www.instituteofplay.com)

programs for all ages and technical abilities. In fact, one of their primary goals is to foster collaboration and an exchange of ideas between students, educators, and professionals. Through gaming the Institute of Play hopes to explore new ways to think, act, and create.

Quest to Learn: New York, New York Scheduled to open in the fall of 2009, Quest to Learn, a school using game-inspired methods to teach traditional and multimedia literacies, is a joint venture between the Transformative Media at Parsons The New School for Design in collaboration with the nonprofit organization New Visions for Public Schools (See http://www.q2l.org/).

Quest to Learn will be a District 2 school, located in Manhattan, and will open with a 6th grade in fall 2009, adding a new grade each year. See http://www.q2l.org/.

This innovative middle and high school, conceptualized by Katie Salen, Director of Graduate Studies in the Digital Design department at Parsons School of Design, redefines the learning paradigm and actively seeks to change the way institutions of learning are conceived of and built by blurring the traditional line between learning and play. It aims to prepare students for a digitally mediated future through a curriculum structured around the creation and execution of alternate reality games. The project will also act as a demonstration and research site for alternative trends in education funded in part by the MacArthur Digital Media and Learning Initiative.

The New York City Museum School The 400 high-school students at the NYC Museum School (http://schools.nyc.gov/SchoolPortals/02/M414/default.htm) spend up to three days a

NYC Museum School: New York, New York

week at a chosen museum (either the American Museum of Natural History, the Metropolitan Museum of Art, the Children's Museum of Manhattan, or the South Street Seaport Museum) studying with specialists and museum educators. Students work on different projects depending on which museum they choose (e.g., geometry and computer animation at the Children's Museum or navigation at the South Street Seaport Museum). At the end of senior year each student shares a thesis-like project on a chosen theme. The NYC Museum School was founded in 1994 by a former Brooklyn Museum assistant director in partnership with a former teacher with the Lab School in New York. It has been featured in the Bill and Melinda Gates Foundation's "High Schools for the New Millennium" report.

The School of the Future: Philadelphia, Pennsylvania The School of the Future in Philadelphia is unique in that it is the first urban high school to be built in a working partnership with a leading software company, Microsoft. The school opened in September 2006 and serves approximately 750 students in a state-of-the-art, high-tech, and "green" facility. Microsoft's Partners in Learning initiative played an integral part in the design and conceptualization of the school, not through a monetary donation (The School of the Future is funded by the School District of Philadelphia), but through the development of new technologies for both teaching and administrative purposes. Among the most innovative, and controversial, of these technologies is a smart card that allows access to digital lockers and that tracks calories consumed during school meals (breakfast and dinner are also served before and after school). Class

The School of the Future: Philadelphia, Pennsylvania. Image of the School of the Future from http://www.phila.k12.pa.us/offices/sof/images/ms_school.jpg

schedules and locations change every day (the goal being to break down our culture's dependency on time and place), and all rooms are designed with flexible floor plans to foster teamwork and project-based learning. Instead of a library and textbooks all students are given a laptop with wireless access to the Interactive Learning Center, the school's hub for interactive educational material. These laptops are linked to SMARTBoards in every classroom and networked so that assignments and notes can be accessed even from home. The building itself is also unique in its holistic approach. Rainwater is caught and repurposed for use in toilets, the roof is covered with vegetation to shield it from ultraviolet rays, panels embedded within the windows capture light and transform it into energy, room settings auto-adjust based on natural lighting and atmospheric conditions, and sensors in all the rooms turn lights on and off depending on whether the space is being used. In short, the School of the Future incorporates many innovations but also has high-tech interactivity that borders on extreme surveillance that makes it a questionable model for future participatory learning initiatives. For more information, see http://www.phila.k12.pa.us/offices/communications/press_releases/2006/09/07/soffacts.html and http://www.microsoft.com/education/SchoolofFuture-.mspx.

Notes

1. The forums took place on February 8, 2007, in Chicago, Illinois; April 21, 2007, at Duke University in Durham, North Carolina, at Electronic Techtonics: Thinking at the Interface, the first international HASTAC conference; and on May 11, 2007, at the University of California's Humanities Research Institute (UCHRI), in Irvine, California.

2. Learning institutions have made great strides in recent years. See, for example, Jason Szep, "Technology Reshapes America's Classrooms," *New York Times*, July 7, 2008. In our forthcoming book, *The Future of Thinking*, we include an extensive "Bibliography: Resources and Models." Yet there is still significant progress to be made. Learning institutions must reexamine their entire structure and approach to learning before they can truly enter the digital age.

3. Bharat Mehra, Cecelia Merkel, and Ann P. Bishop, "The Internet for Empowerment of Minority and Marginalized Users," *New Media and Society* 6 (2004): 781–802. See also the essays collected in *Civic Life Online: Learning How Digital Media Can Engage Youth*, ed. W. Lance Bennett (Cambridge, MA: MIT Press, 2008).

4. According to the *Guinness Book of World Records*, the title of "oldest" is a matter of dispute but, generally, the order is accepted as: University of Al-Karaouine, in Fes, Morocco (859); Al-Azhar University in Cairo,

Egypt (975); the University of Bologna, Italy (1088); the University of Paris (1150); and Oxford (1167).

5. For an excellent discussion of different access to participation, or what is commonly known as the "digital divide," from a transnational perspective, see Terry Flew, *New Media: An Introduction* (Melbourne: Oxford University Press, 2008). In the People's Republic of China, for example, only ten percent of the population has access to the Internet at present, and virtually all communications on the Internet are under surveillance by the government, an issue of both access and censorship. In the United States, Mehra, Merkel, and Bishop, "The Internet for Empowerment of Minority and Marginalized Users," 782, discuss the roles of educational level, socioeconomic status, income, and race as factors contributing to and also influenced by access to digital technologies.

6. "Wikipedia," on Wikipedia, http://en.wikipedia.org/wiki/Wikipedia:About (March 15, 2008).

7. John Palfrey and Urs Gasser, in *Born Digital: Understanding the First Generation of Digital Natives* (New York: Basic Books, 2008), powerfully make the case for the invisibility of many of these issues to many of those (including the vast majority of students entering college today) who have been raised in a world where digital, participatory learning exists. This is in no way to diminish the fact of the digital divide but to emphasize an epistemological divide separating those who grew up participating in digital culture and those who have learned it in adulthood.

8. Cornelia Dean, "If You Have a Problem, Ask Everyone," *New York Times,* July 22, 2008.

9. The initial posting of the draft manuscript on the Institute for the Future of the Book's Web site in January 2007 amassed over 350 registrants. It has since changed considerably to take the comments and suggestions of these registrants into consideration. All comments through March 2008 have been taken into consideration in this publication.

10. For full discussion of authorship in participatory learning (which uses this project as one of its examples) and how participatory learning practices change the stated and unstated premises of peer review, see Cathy N. Davidson, "Humanites 2.0: Promise, Perils, Predictions," *Publications of the Modern Language Association (PMLA)* 123, no. 3 (May 2008): 707–717.

11. An excellent example of an interactive hybrid is the multiple publication sites for the proceedings of our first HASTAC conference (May 2007). *Electronic Techtonics: Thinking at the Interface,* edited by Erin Ennis, Zoë Marie Jones, Paolo Mangiafico, Mark Olson, Jennifer Rhee, Mitali Routh, Jonathan E. Tarr, and Brett Walters, was published under Creative Commons licensing by Lulu, an open source venture founded by Red Hat CEO Bob Young. The book is available for purchasing as a printed volume or by free digital download. Additionally, a multimedia version is available on the HASTAC Web site (www.hastac.org), and edited talks from the conference appear on the HASTAC YouTube Channel (http://www.youtube.com/user/video4hastac). Finally, the interactive data visualization experiment collaboratively produced for the conference has contributed to a nonprofit research Web site, SparkIP (http://www.sparkip.com), which also has an online for-profit component. All of these various forms of content creation constitute "publishing" in the digital age.

12. For more information on the history of the book, see Cathy N. Davidson, ed., *Reading in America: Literature and Social History* (Baltimore: Johns Hopkins University Press, 1989).

13. Robert Darnton, "The Library in the New Age," *New York Review of Books* 55, no. 10 (June 12, 2008).

14. Comment by Wheat on the Web site for the Institute for the Future of the Book, August 6, 2007 (http://www.futureofthebook.org/HASTAC/learningreport/i-overview/).

15. See Lawrence Grossberg, *Caught in the Crossfire: Kids, Politics, and America's Future* (Boulder, CO: Paradigm Publishers, 2005) for a tren-

chant analysis of ways that class and race factor into the lives of youth and merge in U.S. national policy and ideology; and Philomena Essed and David Theo Goldberg, eds., *Race Critical Theories* (London: Blackwell, 2002), for further discussion of how racism inflects these issues. See also Christopher Newfield, *Unmaking the Public University: The Forty-Year Assault on the Middle Class* (Cambridge, MA: Harvard University Press, 2008).

16. See The Education Trust, "Getting Honest about Grad Rates: Too Many States Hide behind False Data," June 23, 2005 (http://www2.edtrust.org/EdTrust/Press+Room/HSGradRate2005.htm).

17. See http://www.americaspromise.org/APAPage.aspx?id=10354. *Cities in Crisis: A Special Analytic Report on High School Graduation,* released April 1, 2008, chaired by Alma J. Powell of America's Promise Alliance and prepared by Editorial Projects in Education Research Center, reveals that "in the metropolitan areas surrounding 35 of the nation's largest cities, graduation rates in urban schools were lower than those in nearby suburban communities. In several instances, the disparity between urban-suburban graduation rates was more than 35 percentage points."

18. See The Prison University Project (http://www.prisonuniversityproject.org/resources.html) and the Correctional Education Facts from the National Institute for Literacy (www.nifl.gov/nifl/facts/correctional.html).

19. See http://www.hrw.org/english/docs/2008/06/06/usdom19035.htm. "U.S.: Prison Rates Hit New High," Human Rights Watch, Washington, DC, June 6, 2008.

20. Former head of the U.S.-based National Telecommunications Infrastructure Administration (NTIA) Larry Irving was among the first to use the term "digital divide" during the Clinton Administration. However, the George W. Bush Administration has focused on growth of access rather than on gaps and divides in its reports, making it very difficult, on a national level, to assess how much of a divide currently exists across socioeconomic levels, within and across races, ethnicities, language and cultural barriers (such as new immigrant communities). For essays that

focus on some of these issues, see Anna Everett, ed., *Learning Race and Ethnicity: Youth and Digital Media* (Cambridge, MA: MIT Press, 2008).

21. John Seely Brown and Douglas Thomas, "The Play of Imagination: Extending the Literary Mind," *Games and Culture* 2 (2007): 149–172.

22. Although many people use the phrase "Net Age" as a shorthand for "Internet Age," we are here using John Seely Brown's particular use of the term to signal both the Internet and networking, the specific combination that O'Reilly calls Web 2.0 and that seems to us a vastly rich model for learning and a specific challenge to most existing forms of learning institutions. See his keynote address, "The Social Life of Learning in the Net Age," presented at the First International HASTAC conference, Electronic Techtonics: Thinking at the Interface, Duke University, April 19, 2007 (http://www.hastac.org/informationyear/ET/JohnSeelyBrown).

23. James Boyle, "A Closed Mind about an Open World," *Financial Times* (August 8, 2006).

24. For an excellent analysis of the pedagogical requirements for a digital age, see Steve Anderson and Anne Balsamo, "A Pedagogy for Original Synners," in *Digital Youth, Innovation, and the Unexpected,* ed. Tara McPherson (Cambridge, MA: MIT Press), 241–259.

25. Since 1994, Alan Liu has been the "weaver," as he says, of The Voice of the Shuttle: Web Page for Humanities Research (http://liu.english.ucsb.edu/the-voice-of-the-shuttle-Web-page-for-humanities-research/).

26. Alan Liu has circulated a useful and levelheaded set of guidelines he issues to students in his undergraduate college classes about consulting Wikipedia for formal coursework purposes. See http://www.english.ucsb.edu/faculty/ayliu/courses/wikipedia-policy.html.

27. Perhaps the best article available on the advantages and, candidly, the shortcomings of Wikipedia as a collaborative knowledge site and, differently, as a reference work is the entry on "Wikipedia" on Wikipedia (http://en.wikipedia.org/wiki/Wikipedia).

28. For an extended discussion of the new models of mind and brain necessary to envision a new, collaborative, horizontal pedagogy, see Cathy N. Davidson's forthcoming *The Rewired Brain: The Deep Structure of Thinking for the Information Age* (to be published by Viking Press in 2010). In more practical terms, E. O. Wilson, the noted biologist, has been leading a major online undertaking in collaboration with others to provide a comprehensive, open source, online catalog of knowledge about every known biological species. It is, as the home page announces, "an ecosystem of Web sites that makes all key information about all life on Earth accessible to anyone, anywhere in the world." Calling for contribution from any concerned person, the project organizers nevertheless have stringent oversight constraints on the quality of contribution, looking to domain experts as content editors. See Encyclopedia of Life (http://www.eol.org/index).

29. For an excellent discussion of the value system implicit in open source culture, see Christopher M. Kelty, *Two Bits: The Cultural Significance of Free Software* (Durham, NC: Duke University Press, 2008). On networked individualism and society, see Barry Wellman, Anabel Quan-Haase, Jeffrey Boase, Wenhong Chen, Keith Hampton, Isabel Isla de Diaz, and Kakuko Miyata, "The Social Affordances of the Internet for Networked Individualism," *Journal of Computer-Mediated Communication* 8, no. 3 (April 2003). Available at: http://jcmc.indiana.edu/vol8/issue3/wellman.html.

30. HASTAC has taken an active role in exploring a variety of electronic publishing forms. In addition to helping to support Kelty's online version of *Two Bits* (as a free download that can be remixed and commented on) and to publishing the first draft of this book on a collaborative writing site, HASTAC has published the proceedings of its first annual conference with Lulu.com, a self-publishing site that allows users to purchase a book or to download for free as well in a multimedia form. The "proceedings" of the second conference combine multimedia (audio-video) as well as multiauthored live blogging of talks, exhibits, and events as an online archive of the event. Each of us is engaged in

ongoing discussions with various academic presses about contemporary electronic publishing initiatives as the future direction of academic publishing.

31. Elizabeth Gudrais, "Unequal America: Causes and Consequences of the Wide—and Growing—Gap between Rich and Poor," *Harvard Magazine* (July–August 2008) (http://harvardmagazine.com/2008/07/unequal-america.html); Claudia Goldin, *The Race between Education and Technology* (Cambridge, MA: Belknap Press, 2008); Bill Readings, *The University in Ruins* (Cambridge, MA: Harvard University Press, 1996); Mark Gibson and Alec McHoul, "Interdisciplinarity," in *A Companion to Cultural Studies*, ed. Toby Miller (Oxford: Basil Blackwell, 2006); David Theo Goldberg, "Enduring Occupations," in *The Threat of Race* (Oxford: Wiley-Blackwell, 2008); and David Theo Goldberg, *The Racial State* (Oxford: Basil Blackwell, 2002).

32. This is the title for the keynote address that John Seely Brown delivered at the first international conference of HASTAC, Electronic Techtonics: Thinking at the Interface, April 19, 2007, at the Nasher Museum of Art at Duke University. A Webcast is available at www.hastac.org. Some schools, including public schools, are just coming online which seek to institutionalize these newly emergent models of networked learning practices.

33. Henry Jenkins, *Confronting the Challenges of Participatory Culture: Media Education for the 21st Century* (New York: New York University Press, 2006), 257.

34. Many of the contributors to the Institute for the Future of the Book Web site used (often cryptic) usernames in order to register and thus could not be identified in terms of their institutional connections. This does not make their comments any less valuable and is instead a natural product of digital collaboration. We attempted to contact users and ask permission to use their real names and institutional affiliations. Where we received no response, we have used the name or pseudonym they used on the IFB site.

Collaborators

Principal Authors

Cathy N. Davidson Duke University, Ruth F. DeVarney Professor of English and John Hope Franklin Humanities Institute Professor of Interdisciplinary Studies, Cofounder of HASTAC

David Theo Goldberg University of California, Irvine, Professor of Comparative Literature and Criminology, Law and Society, Director of the University of California Humanities Research Institute (UCHRI), Cofounder of HASTAC

Editorial and Research Consultant

Zoë Marie Jones Duke University, Department of Art, Art History and Visual Studies

February 8, 2007, Forum in Chicago, Illinois

James Chandler University of Chicago, Barbara E. & Richard J. Franke Distinguished Service Professor, Department of English, Director, Franke Institute for the Humanities

John Cheng Northwestern University, Lecturer and Acting Director, Asian American Studies Program

Allison Clark University of Illinois at Urbana-Champaign, Associate Director, Seedbed Initiative for Transdomain Creativity

S. Hollis Clayson Northwestern University, Professor of Art History, Bergen Evans Professor in the Humanities, Director of the Alice Kaplan Institute for the Humanities

Noshir Contractor University of Illinois at Urbana-Champaign, Professor, Department of Speech Communication, Department of Psychology, Coordinated Science Laboratory, Research Affiliate of the Beckman Institute for Advanced Science and Technology, Director of the Science of Networks in Communities (SONIC) Group at the National Center for Supercomputing Applications (NCSA), and Codirector of the Age of Networks Initiative at the Center for Advanced Study at the UIUC

Dilip P. Goankar Northwestern University, Professor of Communication Studies, Codirector of the Center for Transcultural Studies

Steve Jones University of Illinois at Chicago, Professor, Department of Communication, Associate Dean, Liberal Arts & Sciences

Julie Thompson Klein Wayne State University, Professor of Humanities, Interdisciplinary Studies Program, Faculty Fellow in the Office of Teaching & Learning and Codirector of the University Library Digital Media Project

Martin Manalansan University of Illinois at Urbana-Champaign, Associate Professor of Anthropology

Lisa Nakamura University of Chicago at Urbana-Champaign, Associate Professor, Asian American Studies, Institute of Communications Research

Mary Beth Rose University of Illinois at Chicago, Professor of English and Gender Studies, Director, Institute for the Humanities

Craig Wacker John D. and Catherine T. MacArthur Foundation, Program Officer in Digital Media and Learning

April 21, 2007, Forum in Durham, North Carolina

Ruzena Bajcsy University of California, Berkeley, Professor, Electrical Engineering and Computer Science, Director Emerita, Center for Information Technology Research in the Interest of Society (CITRIS)

Anne Balsamo University of Southern California, Professor, Interactive Media and Gender Studies in the School of Cinematic Arts, and of Communications in the Annenberg School of Communications

Allison Clark University of Illinois at Urbana-Champaign, Seedbed Initiative for Transdomain Creativity

Kevin Franklin Executive Director, University of Illinois Institute for Computing in the Humanities, Arts, and Social Sciences (CHASS), Senior Research Scientist for the National Center for Supercomputing Applications (NCSA)

Daniel Herwitz University of Michigan, Mary Fair Croushore Professor of Humanities, Director, Institute for the Humanities

Julie Thompson Klein Wayne State University, Professor of Humanities, Interdisciplinary Studies Program, Faculty Fellow in the Office of Teaching & Learning and Codirector of the University Library Digital Media Project

Henry Lowood Stanford University Libraries, Curator for History of Science, Technology, and Germanic Collections

Thomas MacCalla National University, Vice President, Executive Director of the National University Community Research Institute (NUCRI) in San Diego

Stephenie McLean Renaissance Computing Institute (RENCI), Director of Education and Outreach

Tara McPherson University of Southern California, Associate Professor, Gender and Critical Studies, School of Cinematic Arts, Founding Editor, *Vectors: Journal of Culture and Technology in a Dynamic Vernacular*

Mark Olson Duke University, Visiting Professor, Department of Art, Art History and Visual Studies; Director, New Media and Information Technologies, John Hope Franklin Center for Interdisciplinary and International Studies

Douglas Thomas University of Southern California, Associate Professor, Annenberg School for Communication, Director of the Thinking Through Technology project, coinvestigator on the Metamorphosis Project

Kathleen Woodward University of Washington, Professor of English, Director, Walter Chapin Simpson Center for the Humanities

May 11, 2007, Forum in Irvine, California

Anne Balsamo University of Southern California, Professor, Interactive Media and Gender Studies in the School of Cinematic Arts, and of Communications in the Annenberg School of Communications

Jean-Francois Blanchette University of California, Los Angeles, Assistant Professor, Department of Information Studies, Graduate School of Education & Information Studies

Tom Boellstorff University of California, Irvine, Associate Professor of Anthropology, Editor-in-Chief, *American Anthropologist*

Allison Clark University of Illinois at Urbana-Champaign, Seedbed Initiative for Transdomain Creativity

Edward Fowler University of California, Irvine, Professor and Chair, East Asian Languages & Literature, Professor, Film & Media Studies

Deniz Göktörk University of California, Berkeley, Associate Professor, Department of German, Cofounder of TRANSIT, the first electronic journal in German studies

Diane Harley University of California, Berkeley, Senior Researcher, Center for Studies in Higher Education, director of the Higher Education in the Digital Age (HEDA) project

Adriene Jenik University of California, San Diego, Associate Professor, Computer and Media Arts, Visual Arts Department

Rosalie Lack University of California, Office of the President, California Digital Library, Digital Special Collections Director

Toby Miller University of California, Riverside, Professor and Chair, Media & Cultural Studies

Christopher Newfield University of California, Santa Barbara, Professor, English Department, Innovation Working Group, Center for Nanotechnology in Society

Vorris Nunley University of California, Riverside, Assistant Professor of English

Mark Poster University of California, Irvine, Professor of History

Todd Presner Associate Professor, German Studies, Chair, Center for Humanities Computing and Director, Hypermedia Berlin, UCLA

Ramesh Srinivasan University of California at Los Angeles, Assistant Professor of Information Studies, Graduate School of Education and Information Studies

William Tomlinson University of California, Irvine, Assistant Professor, Informatics Department, Bren School of Information and Computer Sciences

K. Wayne Yang University of California, San Diego, Assistant Professor of Ethnic Studies

Institute for the Future of the Book[34]

Christine Alfano Stanford University, Lecturer, Program in Writing and Rhetoric and the Department of English

Craig Avery

Anne Balsamo University of Southern California, Professor, Interactive Media and Gender Studies in the School of Cinematic Arts, and of Communications in the Annenberg School of Communications

Mechelle Marie De Craene Florida State University, Department of Art Education and K–12 Teacher (Special Ed./Gifted Ed.), Founder HASTAC on Ning

Kevin Guidry Sewanee: The University of the South, MacArthur Information Technology Fellow

Steve Jones University of Illinois at Chicago, Professor, Department of Communication, Associate Dean, Liberal Arts and Sciences

Becky Kinney University of Delaware, Instructional Programmer, User Services

LAC

Edward Lamoureux Bradley University, Associate Professor, Multimedia Program and Department of Communication, Codirector, New Media Center

Eileen McMahon University of Massachusetts, Boston, Senior Instructional Designer, Communication Studies

Jason Mittell Middlebury College, Associate Professor, American Studies and Film & Media Culture

rcsha

Alex Reid

Michael Roy University of Illinois at Urbana-Champaign, Seedbed Initiative for Transdomain Creativity, Wesleyan University, Director of Academic Computing Services & Digital Library Projects, Founding Editor, Academic Commons

K. G. Schneider University of Illinois at Urbana-Champaign, Seedbed Initiative for Transdomain Creativity, Blogger, Free Range Librarian

Patricia Seed University of California, Irvine, Professor, History of Science and Technology

Trevor Shaw *MultiMedia & Internet@Schools* magazine

David Silver University of San Francisco, Assistant Professor, Department of Media Studies

Bruce Simon State University of New York, Fredonia, Associate Professor of English

Tpabeles

Wheat

Ben Vershbow Editorial Director, Institute for the Future of the Book

Sarita Yardi Georgia Institute of Technology, PhD Student, Human-Centered Computing program

Additional Scholarly Contributions

Anne Allison Duke University, Chair, Department of Cultural Anthropology

Richard Cherwitz University of Texas at Austin, Professor, Department of Communication Studies and Department of Rhetoric and Writing, Founder and Director, Intellectual Entrepreneurship Consortium (IE)

Jonathon Cummings Duke University, Associate Professor of Management, Fuqua School of Business

Diane Favro University of California, Los Angeles, Professor, Department of Architectural History, Director, Center for Experimental Technologies

Carol Hughes University of California, Irvine, Associate University Librarian for Public Services

Alice Kaplan Duke University, Professor, Department of Romance Studies, Literature, and History

Robin Kirk Duke University, Director, Duke Human Rights Center

Timothy Lenoir Duke University, Kimberly Jenkins Professor of New Technologies and Society, Director, Information Sciences + Information Studies Program (ISIS)

Richard Lucic Duke University, Associate Department Chair and Associate Professor of the Practice of Computer Science, Director of External Relations, Information Sciences + Information Studies Program (ISIS), Curriculum Director of the Department of Computer Science

Robert Nideffer University of California, Irvine, Professor of Art, Codirector, Arts, Computation, and Engineering (ACE) Program, and Director of Center for Gaming and Game Culture

Simon Penny University of California, Irvine, Professor of Art and Engineering, Codirector of Arts, Computation, and Engineering (ACE) Program

Kavita Philip University of California, Irvine, Associate Professor of Women's Studies, Anthropology, and Arts, Computation, and Engineering (ACE) Program

Todd Presner University of California, Los Angeles, Associate Professor, German Studies, Chair, Center for Humanities Computing, and Director, Hypermedia Berlin

Ken Rogerson Duke University, Lecturer, Sanford Institute of Public Policy

John Taormino Duke University, Director, Visual Resources Center

Additional Online Forum Contributors (via www.hastac.org)

Mechelle Marie De Craene Florida State University, Department of Art Education and K–12 Teacher (Special Ed./Gifted Ed.), Founder HASTAC on Ning

David Harris University of São Paulo, PhD Student, Director, Global Lives Project

Kenneth R. Jolls Iowa State University, Professor of Chemical and Biological Engineering

Julie Thompson Klein Wayne State University, Professor of Humanities, Interdisciplinary Studies Program, Faculty Fellow in the Office of Teaching & Learning and Codirector of the University Library Digital Media Project

Michael Roy Wesleyan University, Director of Academic Computing Services & Digital Library Projects; Founding Editor, *Academic Commons*